Feryal Rada

Practical Clinical Chemistry

Feryal Rada

Practical Clinical Chemistry

Clinical Chemistry

Noor Publishing

Imprint
Any brand names and product names mentioned in this book are subject to trademark, brand or patent protection and are trademarks or registered trademarks of their respective holders. The use of brand names, product names, common names, trade names, product descriptions etc. even without a particular marking in this work is in no way to be construed to mean that such names may be regarded as unrestricted in respect of trademark and brand protection legislation and could thus be used by anyone.

Cover image: www.ingimage.com

Publisher:
Noor Publishing
is a trademark of
Dodo Books Indian Ocean Ltd. and OmniScriptum S.R.L publishing group

120 High Road, East Finchley, London, N2 9ED, United Kingdom
Str. Armeneasca 28/1, office 1, Chisinau MD-2012, Republic of Moldova, Europe
Printed at: see last page
ISBN: 978-620-7-47900-9

List of Contents

1. INTRODUCTION

Clinical chemistry is a quantitative science. It is concerned with measurement of amounts of biologically important substances (called analytes) in body fluids. The methods used to measure these substances are carefully designed to provide accurate concentrations. The results that are obtained from such measurements are compared to reference intervals or Medical Decision Level (MDL) to provide diagnostic and clinical meaning for the values.

Biologic Specimens: Blood is the most common biologic fluid collected for clinical laboratory testing. It is usually drawn from a vein (in the arm) directly into an evacuated tube. Typically a tube will hold about 5 mL of blood – enough to perform many clinical chemistry tests.

Occasionally, when collection of blood from a vein is difficult, a sample of capillary blood may be collected by pricking the skin and collecting several drops of blood from the puncture site. An example is the use of heelstick blood for testing of newborns.

Phlebotomy: is the process of drawing a blood sample from a blood vessel. For clinical chemistry testing blood is usually drawn from a vein, typically a vein in the arm or back of the hand. Collecting blood from a vein is called **venipuncture**. The medical professional drawing the blood sample is called a **phlebotomist**.

Blood is the most commonly used specimen for testing in the clinical laboratory. Blood consists of two main parts – a fluid portion (called plasma, which contains the dissolved ions and molecules) and a cellular portion (the red blood cells, white blood cells and platelets). Most clinical chemistry analytes are found in the plasma. Part of the preparation of blood for testing these analytes involves removing the cells. This is done by centrifugation of the sample to pack the blood cells in the bottom of the collection tube and allow removal of the liquid portion for testing.

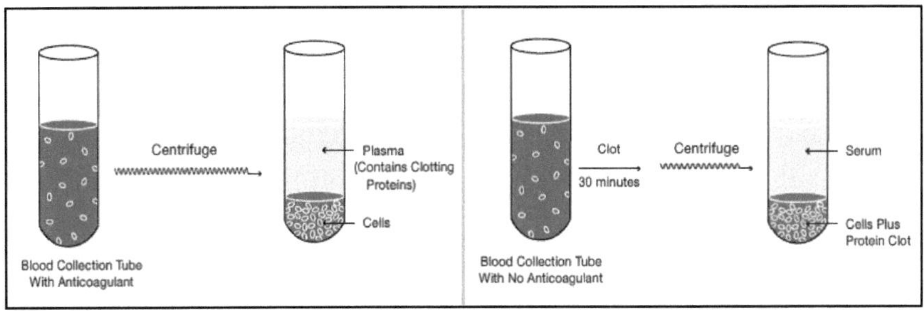

If a blood sample is collected in a tube containing an additive that prevents the blood from clotting (called an anticoagulant), the fluid portion of the blood is called plasma.

If the blood is collected in a tube with no anticoagulant, the blood will form a clot. A clot is a gelatinous semisolid composed of cross-linked protein that is formed in a multi-step process referred to as the clotting cascade.

Upon centrifugation the clot descends to the bottom of the tube along with the cells. The resultant liquid above the cells and clot is called serum. Serum contains all the components of plasma except the clotting proteins, which are consumed in the cascade of reactions that form the blood clot. Some clinical chemistry tests are best performed using plasma, others are best performed using serum.

Tubes used to collect blood have color-coded caps that signal what additives are present in the tube. Additives may be anticoagulants to allow preparation of plasma or may be substances included to protect analytes from chemical or metabolic breakdown.

Note: Certain types of anticoagulants may be incompatible with some kinds of tests. For e.g. Ethylene diamine tetra acetate (EDTA) is an anticoagulant that inhibits the clotting of blood by sequestering (chelating) calcium ions that are necessary components of clotting reactions.

However, samples of plasma collected using EDTA tubes are generally unsuitable for measurement of calcium and for any test method that involves a reaction step depending on availability of calcium.

Sodium fluoride: Preservative and anticoagulant for blood glucose. It inhibits the enzyme enolase and stop further degradation of glucose to pyruvate.

Stopper color	Additive	Used
Red	None	Collecting serum sample
Purple	EDTA	**whole blood Hematology determination**
Pink	EDTA	**Crossmatch EDTA tube ,used for blood transfusion**
Blue	Sodium citrate	PT & PTT tube (coagulation tests)
Black	Sodium citrate **Buffer**	ESR tube **(Erythrocyte Sedimentation Rate)**
Gray	Sodium fluoride	Blood sugar tube
Green	Lithium heparin	Collecting plasma sample
Gold	None	Serum separator tube (SST) contains a gel at the bottom to separate blood from serum

Effects of hemolysis:

1. Hemoglobin color interferes with colorimetric measurement.
2. Glucose decrease due to glycolysis by RBCs enzymes
3. Bilirubin decrease due to color interference
4. Increased potassium, LDH and GOT activity due to leakage from RBCs
5. Increased phosphorus as organic esters in the cell is hydrolyzed

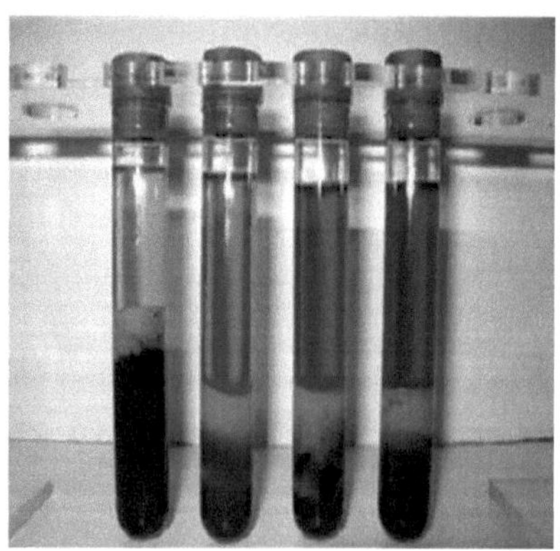

__Photometry:__ is used for measurement of light by a **photodetector**. The light may be absorbed by a substance dissolved in solution (absorbance), the light may be scattered or refracted by particles suspended in solution (turbidimetry or nephelometry),or the light may be emitted from a substance that absorbs light at one wavelength and emits light at another wavelength (fluorescence).

__Colorimetry (absorption photometry):__ Colorimetry is a machine used to measure the amount of light absorbed by the colored substance. The substance to be estimated should be either colored or capable of forming chromogens by the addition of reagents. Colored solution has the property of absorbing certain wave lengths of light and transmitting the others. The absorbance measured based on Beer`s Lambert`s law.

Specific wavelengths of light are chosen for each analysis based on the properties of the substance being measured. A typical **light source (lamp)** generates a broad range of wavelengths of light. A **visible lamp** produces light of wavelengths from 400 nm (violet light) to 700 nm (red light). An **ultraviolet lamp** produces light of wavelengths from about 200 to 400 nm.

To select the desired wavelength from the spectrum of light produced by the light source, a device called a **monochromator or filters** are used. A prism disperses light and monochrometer allows selection of a narrow band of wavelengths to be directed through the sample cuvette and absorb light of unwanted wavelength.

The solution absorbs part of the light and the remaining light is allowed to fall on the photocells which convert light into an electrical signal which is directly proportional to the intensity of the light falling on the detector. These signals are measured by a galvanometer and read as absorbance.

Cuvette: Cell made of optically transparent material that contains solutions for analysis by optical methods.

The selective absorbance of certain wavelengths of light from the spectrum of white light gives the solution its color.

For e.g, a solution containing hemoglobin appears red because light in the green range of the spectrum (wavelengths of 500-600 nm) is selectively absorbed (removed from the white spectrum).

Measuring the decrease in green light that occurs upon passing through the solution, gives an indication of the amount of hemoglobin present. Compounds that have no visible color often absorb light in the ultraviolet region and this absorbance can be used in the same way as absorbance of visible light.

The specific wavelength of light chosen is based on the absorption properties of the compound being measured. As the amount of a substance in solution increases, the relative amount of light that passes through solution and reaches the detector decreases. The decrease in light is termed absorbance.

Absorbance : Absorbance is the measure of the quantity of light that a sample absorb (neither transmits nor reflects) and is proportional to the concentration of a substance in a solution.

The absorbance of a solution depends on:
- Nature of substance
- Concentration of substance
- Wavelength of light
- Path of light

Optical density (OD): The O.D. is used the same as absorbance and directly proportional to the concentration of the colored compound.
As Concentration (C) increases, light **Absorption (A)** increases, linearly.

Transmittance: The transmittance of a sample is the ratio of the intensity of the light that has passed through the sample to the intensity of the light when it entered the sample and displayed as a percentage (%T). As Concentration (C) increases, light **Transmission (%T)** decreases.

Beer's - Lambert's Law : For a given method the coefficient of absorbance (a) and the length of the cell (b) are constants, so the change in absorbance (A) is directly proportional to the concentration (C).

$$A = a * b * C$$

A= absorbance
a or (ε) = coefficient of absorbance or the extinction coefficient (unique property of the substance)
b= the path length of the cuvette containing the solution =1mm
C = concentration of the substance in solution, expressed in (mol /L)

For same compound used standard (known concentration) to find the unknown concentration of analyte (u) after finding the absorbance of standard (s) and analyte then apply this equation:

$$Cu = Cs * Au / As$$

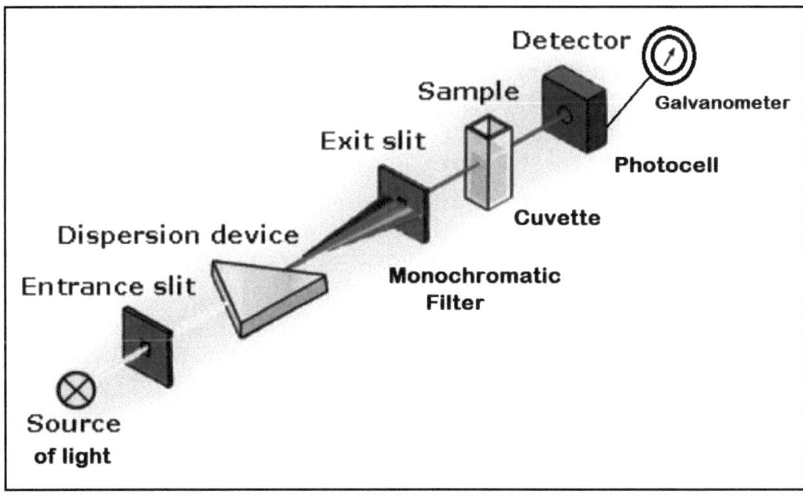

Colorimeters

2. LABORATORY QUALITY CONTROL

Quality Control: is the process of detecting errors

Quality Assurance: Are the systems or procedures in place to avoid errors occurring.

Error Classification:
A) **Pre-analytical** :   errors before the sample reaches the laboratory

B) **Analytical** : errors during the analysis of the sample

C) **Post-analytical** : errors occurring after the analysis

A) Pre - Analytical Errors : involved

1. Improper preparation of the patient:-patient fasting (glucose test), stress and anxiety (urinary protein),drug intake ,position (posture) of patient.

2. Improper collection of the blood sample: sample haemolysis (LDH, potassium or inorganic phosphate), insufficient sample volume (unable to carry out all requested tests), collection timing (24 hour urine), blood stasis, tornicate used < 2 min, vein site.

3. Incorrect specimen container: serum or plasma, fluoride tubes for glucose (to inhibit glycolysis),EDTA unsuitable anti-coagulant for calcium.

4. Improper sample transport for measuring (blood gas, lactic acid, ammonia) : blood sample must put in ice bath and the time between sample collection and laboratory receipt should not exceed 45 min.

5. Incorrect specimen storage: sample left overnight at room temperature (falsely elevated K, Pi and red cell enzymes), delay in sample delivery (falsely lowered levels of unstable analytes).

-Blood stasis due to prolonged tornicate used lead to:

1-Diffusion of the plasma water into interstitial space and the serum or plasma sample obtained will be concentrated (protein,Ca^{+2} ion, thyroxine)

2- Slow of blood flow lead to : increase blood PH ,decrease PO_2, increase PCO_2, increase lactate production due to anaerobic metabolism.

B)-Analytical Errors :

1. The sample: labeling, preparation (centrifugation /aspiration), storage temperature (short -term refrigeration, medium term freezing at -20 ^0C, long term freezing at-80 ^0C), correct test selection (Laboratory Information Management System (LIMS))

2. Glassware / pipettes / balances: used incorrectly, contaminated, poorly calibrated, reuse of pipette tips.

3. Reagents / calibrators / controls: Poor quality, inappropriate storage, correct, temperature, badly maintained fridges or freezers, stability ,shelf-life / working reagent, incorrect preparation. Calculation errors: incorrect factor / wrong calibration values.

4. The application: incorrect analytical procedures, poorly optimized instrument settings, lack of training, the human factor: tiredness / carelessness / stress.

5. The instrument: operational limitations (temperature control/read times/ mixing/ carry-over-lack of maintenance), worn tubing / optics / cuvettes / probes.

C)-Post - Analytical Errors :

1. The correct delivery of the correct report on the correct patient to the correct doctor.

2. How the Clinician interprets the data to the full benefit of the patient.

- Quality control is of two kinds:

1-Internal quality control, the procedure making use of results of only one laboratory for quality control.

2-External quality control in which the results of several laboratories which analyse the same sample(s) are used.

-Two major types of errors may occur in a laboratory:

1-Random errors that arise due to inadequate control on pre-analytical variables, patient identity, sample labelling, sample collection, handling and transport, measuring devices etc.

2-Systemic errors that occur due to inadequate control on analytical variables; e.g. due to error in calibration, impure calibration material, unstable/ deteriorated calibrators, unstable reagent blanks etc.

Quality control involves consideration of a reliable analytical method. Reliability of the selected method is determined by its accuracy, precision, specificity, and sensitivity; with major emphasis of QC being laid on monitoring the precision and accuracy of the performance of analytical methods and take corrective action if needed.

Specificity : The ability of the method to measure solely the analyte it is required to measure. A lack of specificity will affect accuracy and lead to falsely elevated values (hormones and drugs) or falsely low values.

Sensitivity: is the ability of an analytical method to detect small quantities of the measured analyte. It is affect both precision and accuracy at the bottom end of the assay range.

Precision: refers to the extent to which repeated determination on an individual specimen vary using a particular technique. It is the reproducibility of an analytical method. (Give the same result when re measure the same sample).

Accuracy: is the closing the mean of the measured values to the actual value. It is dependent on the methodology used.

3. ESTIMATION OF BLOOD GLUCOSE

Principle: glucose is oxidized by glucose oxidase (GOD) to give gluconic acid and hydrogen peroxide, which is then broken down by peroxidase to water, and oxygen that oxidizes phenol that combines with 4- amino-phenazone to give a red colored complex. The intensity of the colored complex is proportional to the concentration of glucose in the test.

-Methods of blood glucose estimation:

A)-Glucose Oxidase Method (GOD):

$$\text{glucose} + O_2 + H_2O \xrightarrow{\text{Glucose oxidase}} \text{gluconic acid} + H_2O_2$$

$$2H_2O_2 + \text{aminophenazone} + \text{phenol} \xrightarrow{\text{Peroxidase}} \underset{\text{color complex}}{\text{quinoneimine}} + 4H_2O$$

B) Hexokinase Method:

$$\text{glucose} + ATP \xrightarrow{\text{Hexokinase}} \text{glucose 6 phosphate} + ADP$$

$$\text{G-6-P} + NAD \longrightarrow 6\text{-phosphogluconate} + NADH + H^+$$

Clinical Elucidations:

A) Hyperglycemia (increase blood glucose) found in:

1. Diabetes mellitus

2. Cushing's syndrome (adrenocortical hyperactivity)

3. Hyperthyroidism

4. Hyperpituitarism (acromegaly, gigantism)

5. Emotional stress (small increase)

6. Infections

7. Intracranial disease (meningitis, tumors)

8. Effect of drug (corticosteroid, esterogen)

B)-Hypoglycemia (decreased blood glucose level) found in:

1. Starvation, Sever exertion

2. Hyperinsulinism (increase dose of insulin in diabetic patient, tumors of pancreas).

3. Hypothyroidism (myxedema, cretinism)

4. Hypopituitarism (simmonds disease)

5. Hypoadrenalism (Addison disease)

6. In children –glycogen storage disease (von Gierkes disease due to G-6-phosphate deficiency).

Procedure :

Reagents	Blank	Standard	Test
working reagents	1 mL	1 mL	1 mL
Serum or plasma	-	-	0.01 mL
Standard (100 mg /dL)	-	-	0.01 mL
Distilled water	0.01 mL	-	-

Mix the contents of the test tubes thoroughly, and place them in a water bath at 37 ^{0}C for 15 min. or at 25 ^{0}C (+5) for 30 min. Measure the optical density (OD) of the test (sample) and the standard against blank at 515 nm (range 500- 530). The final color complex is stable for more than 2 hours at room temperature.

Calculation:

Glucose conc. (mg/dL)= C (standard) * A (sample) / A (standard)

<u>Blood glucose (normal subject):</u>
Fasting < 100 mg/dL ,
Random (1hr.) < 140 mg/dL ,
Post prandial (2hr.) < 120 mg/dL

4. GLUCOSE TOLERANCE TEST (GTT)

Glucose Tolerance is defined as the capacity of the body to tolerate an extra load of glucose. Normally the blood glucose level remains relatively constant the fasting being 63 -100 mg % which returns to normal within 2 hours. The definitive diagnostic test for DM is the G.T.T.

Procedure of Oral GTT : After an overnight fast of 12-16 hr. Fasting blood sample is taken. Then 75 gm. of glucose dissolved in 250 – 300 ml of water is given orally. Blood samples are collected at 30 min interval for 2-3 hours but 2 hours sample is most important for interpretation of result according to WHO criteria. Corresponding urine samples can also be collected and presence of reducing sugar tested by Benedicts qualitative test. Blood sugar in each sample is estimated. A curve between time and blood glucose concentration, is plotted.

Precautions :
1) The patient should be on a diet of 300 gm. of carbohydrate /day for at least last 3 days.

2) Fasting should not be less than 10 hr. and not more than 16 hr. Only water is permitted after dinner.

3) The patient should not be taking drags that affect carbohydrate metabolism.

4) During the test patient activity should be normal (mild to moderate) and he should abstain from smoking. The patient should be at rest mentally.

Factors affecting GTT:
1) Starvation/Ingestion of high fat diet/ Exercise.

2) Pregnancy-tolerance is decreased.

3) Physiologically decreased tolerance with age.

4) Illness-stress cause's decreased tolerance, so patient recovered from surgery, burns or child birth should be allowed 2 weeks' time before the test is carried out.

5) Endocrine disorders , Liver diseases

6) Drugs-certain drugs must be withdrawn before the test e.g. Oral contraceptives, thiazide diuretics, insulin, oral hypoglycemic agents, salicylates etc.

Clinical elucidations : The following important type of response are seen commonly:

a) ***Normal Response***: Fasting blood sugar is normal. After 1 hr. level rises, but remain below the renal threshold of 180 %. It returns to normal fasting level within 2 hr.

b) ***Diabetic curve***: Fasting level are 7.8 mmol/L (140 mg/ dL) and 2 hour venous blood glucose of 200 mg/dL (11.1 mmol/L) or more are diagnostic of diabetes. Glycosuria is usually seen.

c) ***Impaired GTT***: two hours values of blood glucose between 140 mg/dL and 200 mg/ dL are not abnormal and must be followed up for DM.

d) ***Renal Glycosuria***: Curve is normal Due to lowered renal threshold; one or more samples of urine contain glucose.

e) ***Lag storage /Alimentary Type***: Fasting blood glucose is normal. Due to rapid absorption, maximum level found at 30 min that crosses 180 mg/dL and hypoglycemic levels may be reached at end of 2 hours.

f) ***Flat curve of glucose tolerance***: the fasting blood glucose level is normal. Throughout the test the level does not vary + 20 mg/dL.

Extended GTT :_In this blood samples are taken for 4-5 hr. after giving glucose to see how the curve behaves below the normal fasting glucose limits. Done in some conditions causing hypoglycemia.

Cortisone Stressed GTT: used for detecting latent Diabetes mellitus.

Intravenous GTT: IV. is done if oral glucose is not tolerated or oral GTT curve is flat. In these cases 20 % glucose as 0.5 g glucose/ Kg body weight is infused. Blood samples collected hourly. Usually peak occurs within 30 min after infusion and returns to normal after 90 min.

Blood glucose (impaired glucose tolerance):
Fasting < 120 mg/ dL or [140 mg/ dL]
Random (1hr.) < 180 mg/ dL or [200 mg/ dL]
Post prandial (2hr.) 120 – 180 mg/ dL

Blood glucose (diabetic patients DM):
Fasting >120 mg/ dL
Random (1hr.) >200 mg/ dL
Post prandial (2hr.) >180 mg/ dL

5. ESTIMATION OF BLOOD UREA

Urea is the end product of protein metabolism in our body. Synthesis of urea occurs in the liver cell from ammonia and CO_2, which are the catabolic end products of amino acids.

Urea is hydrolyzed by Urease into ammonia and carbon dioxide. The ammonia generated reacts with alkaline hypochlorite and sodium salicylate in presence of sodium nitroprusside as coupling agent to yield a green cromophore. The intensity of the color formed is proportional to the concentration of urea in the sample.

-Normal value = 15 - 40 mg/ dL (2.5 – 6.6 mmol/L)

$$NH_3-\overset{\overset{O}{\|}}{C}-NH_3 + H_2O \xrightarrow{\text{UREASE}} 2NH_3 + CO_2$$

Urea

$$NH_4^+ + \text{Salycilate} + NaClO \xrightarrow[\text{OH}^-]{\text{NITROPRUSSIDE}} \text{Indophenol} + NaCl$$

Normal blood urea in adults is 15-50 mg % when expressed as blood urea nitrogen (BUN) it is 7-25 mg %. It is somewhat higher in men than women and there is gradual rise with age. The urea concentration varies with the amount of protein in the diet.

Serum urea is lower in pregnancy probably due to hoemodilution, usually between 15-20 mg /100 ml. Increase in urea is significant as a measure of renal function. The causes may be:

A)-*Pre Renal*: When there is reduced plasma volume it leads to decreased renal blood flow and hence GFR leading to urea retention. seen in

1-Reduced Plasma Volume :-

- Acute intestinal obstruction ,

- Prolonged vomiting,

- Severe diarrhoea,

- Ulcerative colitis.
- Pyloric stenosis.
- Diabetic Keloacidosis.
- Shocks, severe burns and haemorrhage.

2- Increased Protein Catabolism
- Fevers
- Thyrotoxicosis
- Cardiac failure

B)-*Renal Disease*: Blood urea is increased in all forms of renal diseases like;
- Acute glomerulonephritis.
- Renal failure
- Malignant hypertension
- Chronic pyclonephritis
- Congenital cystic kidneys

C)-*Post Renal Cause*: Due to obstruction to flow of urine there is retention and so reduction in effective filtration pressure at the glomeruli.
- Enlargement of prostrate.
- Stones in urinary tract.

Procedure:

-Reagent Preparation:

1. Bring reagents and samples to room temperature and then mix one volume of R1 with 24 volumes of R2, stable for 4 weeks at 2-8 °C

2. Pipette into test tubes.

Tubes	Blank	Sample	Standard
Working reagent	1 mL	1 mL	1 mL
Sample	-	10 µL	-
Standard	-	-	10 µL
Mix and incubate for 5 min. at 37 ^{0}C or for 10 min. at room temperature (16-25 ^{0}C).			
Reagent R3	1 mL	1 mL	1 mL

3. Mix thoroughly and incubate the tubes for 5 min. at 37^0 C or for 10 min. at room temperature (16-25) ^{0}C.

4. Read the absorbance (A) of the samples and the standard at 600 nm against the reagent blank. **Note:** the color is stable for at least 2 hr. protected from light.

Calculations:

Serum ,Plasma

$$\text{Urea level (mg/dL)} = \frac{\text{A sample}}{\text{A standard}} \times \text{C standard}$$

6. ESTIMATION OF BLOOD CREATININE

Creatine is synthesized in the liver, kidneys and pancreas, and is transported to its sites of usage, principally muscle and brain. About 1-2 % of the total muscle creatine pool is converted daily to creatinine through the spontaneous, non-enzymatic loss of water. Creatinine is an end-product of nitrogen metabolism, and as such undergoes no further metabolism, but is excreted in the urine.

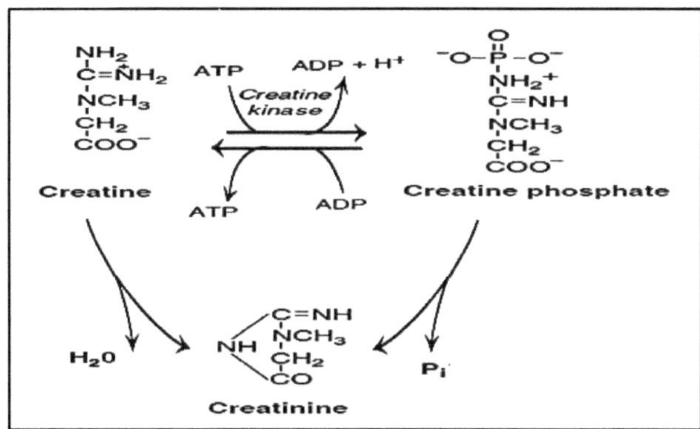

Principle of assay is Photometric Colorimetric Test (Jaffe-Reaction) . Creatinine forms in alkaline solution an orange-red colored complex with picric acid. The absorbance of this complex is proportional to the creatinine concentration in the sample.

Creatinine + Picric acid   Creatinine - Picrate complex

Reference Values : Men = 0.6 – 1.2 mg/ dL , Women=0.5 – 0.9 mg/ dL

Clinical elucidation :

A)-Increased level of creatinine in plasma is found in:

1-Renal disease

2-Congestive heart failure

3-Obstruction of urinary tract

B)-Decreased creatinine clearance is a sensitive indicator of reduced glomerular filtration rate.

***Creatinine Clearance*:** Method used for assessing glomerular function . creatinine is filtered fully by the glomerulus but partially reabsorbed by the tubules of the kidney.

Creatinine Clearance = U x V/ P

U : Concentration of creatinine in urine (mg/100 mL)
P : Plasma creatinine concentration (mg/100 mL)
V: Urine flow in one minute (ml/min)

Normal level : Male: 85 -125 ml/min ,   Female : 80 -115 ml/min

Procedure:

Pipette into test tubes.

Tubes	Blank	Sample	Standard
Sample	-	100 µL	-
Standard	-	-	100 µL
Working reagent	1 mL	1 mL	1 mL
Mix and start the stopwatch. after 30 second read the absorbance A1 at 492 nm (490-510nm) .after 2 min read the absorbance A2 . A2-A1=ΔA sample or ΔA standard			

Calculations: (Serum, Plasma)

Creatinine conc. = C standard x A sample /A standard

C standard = 2 mg/ dL = 176.8 µmol/L

Creatinine conc. (mg/ dL) = 2 x A sample /A standard

Creatinine conc. (µmol/L) = 176.8 x A sample /A standard

7. ESTIMATION OF BLOOD TOTAL PROTEIN

Serum is a straw colored fluid separated when the blood has clotted and it lacks the clotting factors such as prothrombin and fibrinogen.

Plasma is the supernatant obtained when an anticoagulant added blood is centrifuged. It contains all the coagulation factors along with other proteins .plasma contains hundreds of different proteins. All the plasma proteins are synthesized in liver except immunoglobulin which are synthesized in the plasma cells.

Total plasma proteins	6 – 8 gm/100 ml (dL)
Albumin	3.5 – 5.8 gm/ dL
Globulins	2.5 – 3.5 gm/ dL
Fibrinogen	200 –400 mg/ dL
Albumin / globulin ratio	1.5 / 1

- ***Methods of fractionation of plasma proteins:***

1- Half saturation with ammonium sulphate.

2- Cohn`s fractionation by precipitating protein at different PH and ethyl alcohol gradient.

3- Electrophoresis.

4- Ultra centrifugation

-Methods of estimation of plasma proteins:

1-Lowery`s method : estimation of tyrosine in protein.

2- Copper sulphate specific gravity method.

3- Biuret method (commonly used method)

-Biuret method (photometric colorimetric test) :

Principle: The assay is based on polypeptide chelation of cupric ion (colored chelate) in strong alkali. Compounds containing two or more peptide bonds ($CO-NH_2$) react with cupric ions (Cu^{2+}) in alkaline solution to produce a complex of reddish-purple color. The intensity of the color (absorbance of complex) is directly proportional to the concentration of total proteins present in the sample.

Clinical elucidation:

A-Plasma protein level increased in:

1-Hemoconcentration : due to dehydration from loss of fluids (vomiting ,diarrhea)

2-Physiological causes: standing position, vigorous exercise, excessive stasis while taking blood

3-Pathological causes: infective disease (tuberculosis), multiple myeloma.

B -Plasma protein level decreased in :

1-Physiological: pregnancy

2-Malnutrition (kwashiorkor): impaired intake of protein

3-Cirrhosis of liver : reduced synthesis

4-Malabsorption: peptic ulcer, enteritis, steatorrhea

5-Excessive loss in urine: nephrotic syndrome (proteinuria / albuminuria)

6-Excessive loss from intestine : protein losing enteropathy

7-Loss from the skin : burns

8-Sever hemorrhage

9-Fever and after injury: increased catabolism

C – Reversal of A:G ratio :

1. Reduction of albumin : (liver disease, nephrotic syndrome)

2. Elevation of globulin: chronic infections.

D–Urine protein level increase in :

Healthy adults normally excrete up to 150 mg/ 24 hr. (day) protein in urine daily (for this amount, less than 30 mg should be albumin). More than 150 mg/ day defined as **proteinuria**. Protein in the urine may be physiological proteinuria or pathological proteinuria.

A. Physiological proteinuria involves:

1. Transient proteinuria: known as functional proteinuria

2. Intermittent proteinuria: known as orthostatic proteinuria

B. Pathological proteinuria involves:

1. Pre renal proteinuria: known as Overflow or overload proteinuria.

2. Renal proteinuria:

   a. Glomerular proteinuria (albuminuria): nephrotic syndrome

   b. Tubular proteinuria: due to Nephrotoxic drugs (e.g.aminoglycosides), Heavy metal (e.g., lead, mercury, cadmium), Poisoning Acute tubular necrosis .

3.Post-renal proteinuria: due to haemorrhage or inflammation within the lower urinary tract or genital tract.

Degree of proteinuria	Amount of protein
Microalbuminuria	30-300 mg/ day
Mild	300-500 mg/ day
Moderate	500-1000 mg/ day
Heavy	1000-3000 mg/ day
Nephrotic range	More than 3500 mg/ day

Procedures

Pipette into test tubes:

Tubes	Reagent blank	Sample	Standard
Sample	-	20 µL	-
Standard	-	-	20 µL
Reagent	1000 µL	1000 µL	1000 µL
Mix, incubate for 10 min.at 25°C.Measure the absorbance at 550 nm of the sample and standard against the reagent blank within 30 min. (ΔA).			

Calculation :

Total Protein conc. = C standard x ΔA sample / ΔA standard

Total Protein conc. (g/dL) = 8 g/dL x ΔA sample / ΔA standard

8. Estimation of Blood Uric Acid

Uric acid is the end product of the metabolism of purines which are adenine, guanine, xanthine and hypoxanthine. Purines are present in the nucleic acid. Uric acid is one of the non-protein nitrogenous substances.in kidney uric acid is fully filtered by the glomerulus but partially reabsorbed in the proximal convoluted tubules .The normal uric acid level in the blood : male (3.5 – 7 mg/ dL), female (2.5- 6.5 mg/ dL). Daily urinary excretion of uric acid is 250 - 750 mg , at body PH, uric acid exists as monosodium urate salt.

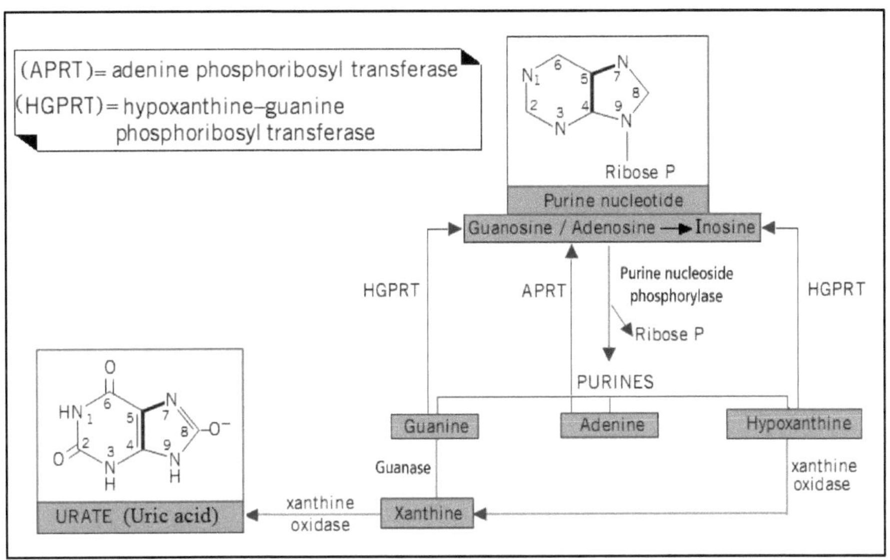

Synthesis of Uric acid

Principle: (colorimetric method),uric acid is converted by uricase to allantoin and hydrogen peroxide ,which under the catalytic influence of peroxidase ,oxidizes 3,5-dichloro-2-hydroxybenzenesulfonic acid and 4-aminophenazone to form a red –violet quinoneimine compound .

$$\text{Uric Acid} + O_2 + 2H_2O \xrightarrow{\text{Uricase}} \text{Allantoin} + CO_2 + 2H_2O_2$$

$$2H_2O_2 + 3,5\text{-dichloro-2-hydroxybenzenesulfonic acid} + 4\text{-aminophenazone}$$

$$\xrightarrow{\text{peroxidase}} \text{N-(4-antipyryl)-3-chloro-5-sulfonate-p- benzo-quinoneimine}$$

Clinical elucidation :

A)-Hyperuricaemia :

1- Gout disease at which the urates are deposited in solid form in and about the joins causing arthritis. Levels are not related to severity of disease.

2-Renal failure due to decreased excretion, blood urea also increased.

3-Conditions of increased turnover of cells as in leukemia, myeloproliferative diseases, pernicious anaemia, chronic hemolytic anaemia.

4-Toxaemia of pregnancy.

5-Diabetes Mellitues.

6-Starvation.

7-Drugs like pyrazinamide, diuretics.

B)-Hypouricaemia :

1- Liver diseases (where the maximum synthesis of uric acid occurs) like cirrhosis or Wilson's disease.

2- Renal disease that decrease renal tubular reabsorption like Fanconi's Syndrome.

3- Drugs-(Uricosuric agent) in large does like salicylates, sulphinpyrazone.

Procedure :

1. Pipette into test tubes.

Tubes	Blank	Sample	Standard
Sample	-	20 µL	-
Standard	-	-	20 µL
Working reagent	1 mL	1 mL	1 mL

2. Mix, and incubate the tubes for 15 min. at 20- 25 ^{0}C

3. Read the absorbance (A) of the samples and the standard at 520 nm against the reagent blank within 30 min.

Calculation:

Uric acid conc. = C standard x A sample /A standard

C standard = 10 mg/dL =595 µ mol/L

Uric acid conc. (mg/dL) = 10 x A sample /A standard

Uric acid conc. (µ mol/L) = 595 x A sample /A standard

9. Estimation of Blood Bilirubin

Bilirubin is the bile pigment formed from degradation of hemoglobin in the reticuloendothelial system. This bilirubin is then conjugated in the liver to form bilirubin glucuronides, which will be secreted in the bile and stored in gallbladder from where it passes into duodenum through common bile duct.

Principle of assay involves the reaction of Bilirubin with diazotised sulphanilic acid (DSA) to form a red azo dye. The absorbance of this dye at 546 nm is directly proportional to the bilirubin concentration in the sample. Water-soluble bilirubin glucuronides react directly with DSA whereas the albumin conjugated indirect bilirubin will only react with DSA in the presence of an accelerator.

(Total bilirubin = direct + indirect bilirubin.)

Sulphanilic acid + sodium nitrite $\longrightarrow$ DSA

Bilirubin + DSA $\longrightarrow$ DIRECT Azobilirubin

Bilirubin + DSA + accelerator $\longrightarrow$ TOTAL Azobilirubin

Clinical elucidation :

Normal serum bilirubin is less than 1 mg/dL, Hyperbilirubinemia of more than 3 mg/dL results into clinical jaundice.

- Classification of jaundice:

1. As per the nature of bile pigment:

   a)-Conjugated hyperbilirubinemia

   b)-Unconjugated hyperbilirubinemia

2. As per the site of lesion:

   a)-Prehepatic ,   b)-Hepatic ,   c)-Post Hepatic

3. As per the pathological causes:

  a)-Hemolytic,

  b)-Hepatocellular

  c)-Obstruction of bile duct (bile flow)

- Types of Hyperbilirubinemia :

A. Unconjugated Hyperbilirubinemia (Hemolytic or Prehepatic Jaundice): caused by increased degradation of RBC s and hemolysis due to:

1- Physiological jaundice of new born particularly in premature infants who not have fully developed microsomal conjugating enzyme system.

2-Incompatible blood transfusion.

3-Ineffective erythropoiesis e.g. Pernicious anemia

4-Rh Incompatibility.

5-Abnormal RBC e.g. in spherocytosis, haemoglobinopathies, G-6-P dehydrogenase deficiency.

6-Congenital nonhemolytic Unconjugated Hyperbilirubinaemia due to impaired glucuronyl transferase activity (Crigler_Najjar Syndrome) in addition to impaired uptake of bilirubin (Gilbert's syndrome).

B. Conjugated Hyperbilirubinaemia :

1- Hepatic Jaundice (viral hepatitis)

2- Post hepatic Jaundice (Obstructive): Caused by gallstones, tumor.

3- Congenital conjugated Hyperbilirubinemia due to defective biliary transport (Dubin- Johnson 's syndrome)

C. Mixed Hyperbilirubinemia: both unconjugated and conjugated bilirubin are increased, due to hepatitis, drugs, poisoning, and alcoholism.

Procedure

▷**For Total Bilirubin**   546nm wavelength

Pipette into cuvettes	Sample blank	Sample
TBR	1000 µl	1000 µl
TNR	---	1 drop *
Mix thoroughly, incubate for 5 min.		
Sample	100 µl	100 µl
Mix, incubate at room temperature for 10 to 30 min. Measure the absorbance of sample against sample blank (ΔA_{546})		

* 1 drop ≈ 40 µl

▷**For Direct Bilirubin**

Pipette into cuvettes	Sample blank	Sample
DBR	1000 µl	1000 µl
DNR	---	1 drop *
Mix thoroughly, add sample within 2 min.		
Sample	100 µl	100 µl
Mix, incubate at room temperature for **exactly 5 min.** Measure the absorbance of sample against sample blank (ΔA_{546})		

* 1 drop ≈ 40 µl

Calculation

Calculate the concentration of total and direct bilirubin by using the factor 13.0.

Bilirubin concentration [mg/dl] = ΔA_{546} × 13.0

[mg/dl] × 17.1 = [µmol/l]

Normal value :

Total serum Bilirubin	0.2 - 1.2 mg/dL
Conjugated (Direct) Bilirubin	0.1 - 0.4 mg/dL
Unconjugated (Indirect) Bilirubin	0.2 - 0.7 mg/dL

10. ESTIMATION OF BLOOD AMINO TRANSFERASE

Alanine amino transferase (ALT or GPT) and *Aspartate amino transferase (AST or GOT)* catalyze the reductive transfer of an amino group from alanine or aspartate ,respectively ,to alpha-ketoglutarate to yield glutamate and pyruvate or oxaloacetate, respectively.

Damaged hepatocytes release their contents including ALT and AST into the extracellular space.The released enzymes ultimately enter into circulation and thereby increase the serum levels of ALT and AST compared to control subjects.

AST is localized in heart, brain, skeletal muscle and liver tissue. ALT is primarily localized to liver, with lower enzymatic activities found in skeletal muscle and heart tissue. Increases in serum ALT and AST enzymatic levels can also arise from extrahepatic injury, particularly to skeletal muscle.

1-The principle of assay is Colorimetric determination of GOT or GPT activity according to the following reactions:

Aspartate + α- ketoglutarate $\xrightarrow{\text{GOT}}$ Oxaloacetate + glutamate

Oxaloacetate +dinitrophenyl hydrazine (DNPH) $\longrightarrow$ color complex

Alanine + α- ketoglutarate $\xrightarrow{\text{GPT}}$ Pyruvate + glutamate

Pyruvate + (DNPH) $\longrightarrow$ color complex

The pyruvate or oxaloacetate formed is measured in its derivative form 2.4-dinitrophenyl-hydrazonc. Range of expected values in serum: GOT, GPT = (0 − 35) IU/L

International Unit (IU): one unit is define as the amount of enzyme that will catalyze under optimum conditions the transformation of one micromole (μmole) of substrate into product per minute.

2-Kinetic method: monitor the speed or rate of reaction, primarily used for measuring enzymes, and some metabolites (urea, creatinine). It is used ultraviolet region for detection.

Calculations

-Reaction rate:

Change in absorbance (Δ A) / minute = [(A3 -A2)+(A2 -A1)] / 2

-**Enzyme activity (U/L) = (Δ A) / minute x Factor**

-Factor used is a constant dependent on :
   -Sample to reagent ratio
   -Wavelength of measurement (UV region 340 nm)
   -Light path

__ALT (GPT)__: (Decrease in A 340 nm directly proportional to ALT level)

Oxoglutarate + Alanine $\xrightarrow{\text{ALT}}$ Glutamate + Pyruvate

Pyruvate + NADH +H$^+$ $\xrightarrow{\text{LDH}}$ Lactate + NAD$^+$

__AST (GOT)__: (Decrease in A 340 nm directly proportional to AST level)

Oxoglutarate + Aspartate $\xrightarrow{\text{AST}}$ Glutamate + Oxaloacetate

Oxaloacetate + + NADH +H$^+$ $\xrightarrow{\text{MDH}}$ Malate + NAD$^+$

__Clinical elucidation__:

__Elevation in the serum levels of AST and ALT__:
A. *Liver disease* : ALT is the more liver specific enzyme although both AST and ALT are raised in liver diseases.

1. Hepatic necrosis, viral hepatitis elevated levels of AST and ALT are found even before the clinical signs and symptoms appear.

20-25 fold elevation are encountered. Peak values are seen between the 7th and 12th day and values return to normal levels by 3rd to 5th week.

2. Extrahepatic cholestasis: moderate increase in the levels of AST and ALT activities

3. Cirrhosis: Levels vary with the status of the cirrhosis. AST activity is higher than ALT activity.

B. *Myocardial infarction*: Elevated AST level begins 3-8 hr. after the onset of the attack and returns to normal within 3-6 days peak is seen 24 hr. after the onset. ALT levels are within normal limits or only marginally increased. ALT increases in liver damage secondary to heart disease.

C. *AST raised levels are seen in*:

1- In vitro hemolysis of the blood samples

2- Skeletal muscles disease and injury.

3- Pulmonary emboli and hypoxia

4- Acute pancreatitis

5- Haemolytic diseases

D. *Others*: Both AST and ALT levels may be elevated after administration of drugs such as opiates, salicylates.

Procedure:

(Reagents)

R1	ALT substrate	TRIS buffer 150 mmol/L (PH=7.3)
		L-alanine 750 mmol/L
		Lactate dehydrogenase > 1350U/L
R2	ALT coenzyme	NADH 1.3 mmol/L
		2-oxoglutarate 75 mmol/L

1. Reagent Preparation_: Mix 4 ml of R1+1 ml of R2 to prepare working reagent, stable for 4 weeks.

2. Pipette into test tubes.

Reaction temperature	37 °C
Working reagent	1 ml
Sample	50 µL
-Incubate for 1 min. and read initial absorbance at 340 nm.	
-Repeat the absorbance reading after 1,2 and 3 min.	
-Calculate the differences between absorbance.	
-Calculate the mean of the results to obtain the average change in absorbance per min.(ΔA/min).	

Calculation:

ALT activity (IU/L) = ΔA/min x 3333

Procedure:

(Reagents)

R1	AST substrate	TRIS buffer 121mmol/L (PH=7.8)
		L-aspartate 362 mmol/L
		Malate dehydrogenase > 460 U/L
R2	AST coenzyme	NADH 1.3 mmol/L
		2-oxoglutarate 75 mmol/L

1. Reagent Preparation_: Mix 4 ml of R1+1 ml of R2 to prepare working reagent, stable for 4 weeks.
2. Pipette into test tubes.

Reaction temperature	37 °C
Working reagent	1 ml
Sample	50 µL
-Incubate for 1 min. and read initial absorbance at 340 nm.	
-Repeat the absorbance reading after 1,2 and 3 min.	
-Calculate the differences between absorbance.	
-Calculate the mean of the results to obtain the average change in absorbance per min.(ΔA/min)	

Calculation:

AST activity (IU/L) = ΔA/min x 3333

11. Estimation of Blood Alkaline Phosphatase

Alkaline phosphatase (ALP) is associated with cell membranes in multiple tissues, particularly hepatocytes. It hydrolyzes mono phosphatases at an alkaline PH. Several alkaline phosphatase isoenzymes (liver, bone , intestinal, kidney, and placental forms) have been identified in humans.

It is primarily a marker of hepatobiliary effects and cholestasis (moderate to marked elevations). In humans with primary biliary cirrhosis, a chronic cholestatic liver disease of unknown etiology, drug-induced cholestasis , tests reveal elevations of serum alkaline phosphatase activity (20-fold) and GGT activity with or without elevated ALT levels.

The principle of assay is Colorimetric determination of alkaline phosphatase activity according to the following reaction:

$$\text{Phenylphosphate} \xrightarrow[\text{PH } 10]{\text{Alkaline phosphatase}} \text{Phenol} + \text{Phosphate}$$

The phenol liberated is measured in the presence of amino-4-antipyrine and potassium ferricyanide.The presence of sodium arsenate in the reagent stops the enzymatic reaction. Range of expected values in serum: Adults: (35 – 100 U/L)

Clinical elucidation:

The normal range of serum alkaline phosphatase activity is 3-13 KA units in adults and is up to about two and a half times greater in children particularly during periods of active growth.

A. Physiological Increase in levels of serum alkaline phosphatase:

1. Children: during periods of bone growth.

2. Puberty.

3. Pregnancy: third trimester-due to the isoenzyme of placental origin.

B. Pathological Increase in levels of serum alkaline phosphatase seen in :

(A)-Bone Diseases: Activity is increased when osteoblasts are more actively laying down osteoid.

1. Rickets

2. Osteomalacia : there is an increase but not as marked as in case of rickets.

3. Paget's disease (Osteitis deformans): The highest levels of serum alkaline phosphatase activity are reported . Values range from 10-25 times the upper limit of normal.

4. Osteogenic disease

5. Healing of bone fractures: Transient elevation seen.

(B)-Liver Disease: The response of the liver to any form of biliary obstruction is synthesis more alkaline phosphatase. The main sites of new enzyme synthesis are the hepatocytes.

Procedure:

Pipette into test tubes.

Sample	20 µL
BUF	1000 µL
Mix ,incubate for 1 min.at the desired temperature 25°C,30°C or 37°C	
SUB	250 µL
Mix,read the absorbance at 405 nm (400- 420 nm) after 1 min. , then read the absorbance again after 1, 2 and 3 min.	
Calculate the mean of the results to obtain the average change in absorbance per min.(ΔA/min)	

Calculation:

Total Alkaline phosphatase activity (IU/L) = ΔA/min x 3433

12. ESTIMATION OF BLOOD GAMMA–GLUTAMYL TRANSFERASE

Gamma-glutamyl transferase (GGT) activity is localized to liver, kidney, and pancreas tissues. GGT has multiple functions including catalytic transfer of the glutamyl residue from glutathione to an acceptor amino acid, with the formation of cysteinyl glycine .

The operation of the cycle involves hydrolysis and resynthesis of glutathione, with consumption of ATP .GGT activity is a marker of hepatobiliary injury, especially cholestatic and biliary effects.

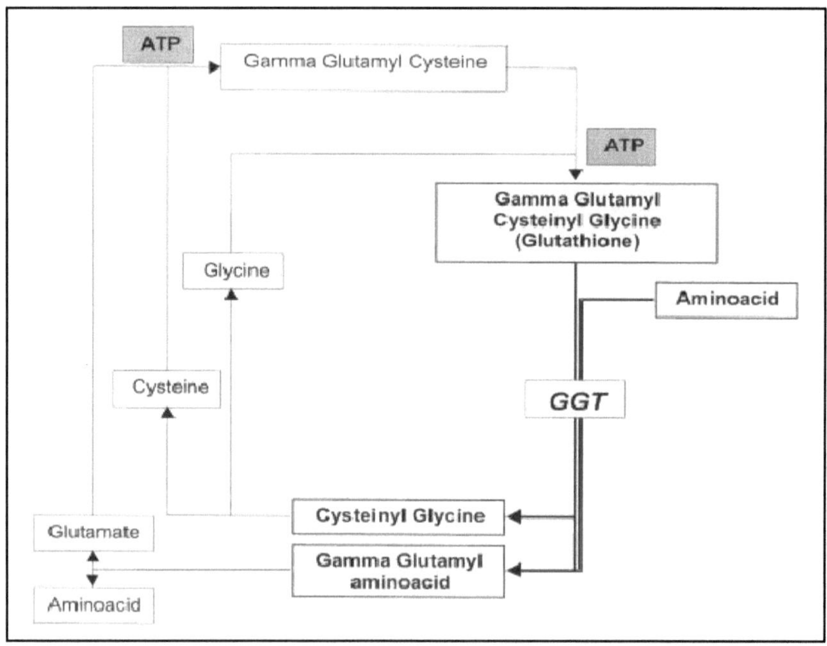

The gamma-glutamyl cycle

Procedure

(Reagents)

Buffer (BUF)	TRIS buffer (PH=8.25),100 mmol/L
	Glycylglycine, 150 mmol/L
Substrate (SUB)	L- γ-glutamyl-3-carboxy-4-nitroanilide

Pipette into test tubes.

Sample	100 µL
BUF	1000 µL
Mix ,incubate for 1 min. at 25°C,30°C or 37°C	
SUB	250 µL
Mix, read the absorbance at 405 nm after 1 min. , then read the absorbance again after 1,2 and 3 min.	
Calculate the mean of the results to obtain the average change in absorbance per min.(ΔA/min)	

Calculation:

γ-Glutamyl transferase activity (IU/L)= ΔA/min x 1606

13. Estimation of Blood Creatine Kinase (CK or CPK)

Creatine kinase (CK) is primarily found in striated muscle, brain and heart tissues. It is responsible for conversion of creatine-to-creatine phosphate in presence of ATP. It is consist of two polypeptide chain B, M therefore 3 isoenzyme are present:

-BB (CK1) : brain, thyroid, other
-MB (CK2) : heart, muscle
-MM (CK3) : heart, skeletal muscle

The determination of CK activity in plasma or serum provides a sensitive marker for detection of skeletal muscle disease e.g. duchenne type muscular dystrophy also it is useful in diagnosis of myocardial infarction and cerebrovascular accidents.

CK increase seen in :-myocardial infarction , muscle dystrophy, after operation , renal failure ,coma, hypothyroidism.

The determination of CK using creatine phosphate and adenosine diphosphate (ADP) as substrates rather than creatine and adenosine triphosphate (ATP) has several advantages in test performance as it allows for a faster reaction rate resulting in greater sensitivity.

Small sample volumes are used and sample blanks are not required. Due to the oxidation of the sulfhydryl groups in the active site of creatine kinase, (very unstable). The enzyme however, is reactivated by the addition of thiol compounds N-acetyl-L-cysteine (NAC), is included in the test for this reason. No sulphate is present in the test system as these alter the reaction kinetics.

UV. Method: It is an optimized standard method according to the recommendations of the Deutsche Gesellschaft fur Klinische Chemie. Increase in A 340 nm is directly proportional to CK. Principle of this method is:

Creatine phosphate + ADP $\xrightarrow{\text{CK}}$ Creatine + ATP

Glucose + ATP $\xrightarrow{\text{HK}}$ Glucose-6-P + ADP

Glucose-6-P + NADP$^+$ $\xrightarrow{\text{G-6-PDH}}$ Gluconate-6-P + NADPH + H$^+$

MANUAL PROCEDURE

Wavelength:		340 nm, Hg 365 nm or Hg 334 nm	

Pipette into cuvette:

	25°C/30°C		37°C	
	Macro	Semi Micro	Macro	Semi Micro
Enzymes/Coenzymes/ Substrate	2.5 ml	1.0 ml	2.5 ml	1.0 ml
Sample	0.1 ml	0.04 ml	0.05 ml	0.02 ml

Mix, incubate for :-

3 min at 25°C , 2 min at 30°C , 1 min at 37°C

Read initial absorbance and start timer simultaneously. Read again after 1, 2 and 3 min.

CALCULATION

To calculate the CK activity use the following formulae:

25°C/30°C	37°C
U/l = 4127 x Δ A 340 nm/min	U/l = 8095 x Δ A 340 nm/min
U/l = 4207 x Δ A 334 nm/min	U/l = 8252 x Δ A 334 nm/min
U/l = 7429 x Δ A 365 nm/min	U/l = 14571 x Δ A 365 nm/min

NORMAL VALUES IN SERUM

	25°C	30°C	37°C
Men	10-80 U/l	15-130 U/l	24-195 U/l
Women	10-70 U/l	15-110 U/l	24-170 U/l

14. Estimation of Blood Total Cholesterol

Cholesterol is a fat-like substance that is found in all body cells. The liver makes all of the cholesterol the body needs to form cell membranes and to make certain hormones. The determination of serum cholesterol is one of the important tools in the diagnosis and classification of lipemia. High blood cholesterol is one of the major risk factors for heart disease.

Principle: (Colorimetric–Enzymatic Method), Free cholesterol and cholesterol released from it's esters after enzymatic hydrolysis are oxidized enzymatically. The indicator quinoneimine is formed from hydrogen peroxide and 4-amino-antipyrine in the presence of phenol and peroxidase.

$$\text{Cholesterol ester} + H_2O \xrightarrow{\text{Cholesterol esterase}} \text{Cholesterol} + \text{fatty acids}$$

$$\text{Cholesterol} + O_2 \xrightarrow{\text{Cholesterol oxidase}} \text{Cholestene-3-one} + H_2O_2$$

$$H_2O_2 + \text{4-Aminoantipyrine} + \text{phenol} \xrightarrow{\text{Peroxidase}} \text{Quinoneimine} + H_2O_2$$

Clinical elucidation:

Serum cholesterol varies from 150-240 mg/100 ml in healthy young adults. The level rises with age and may go up to 300 mg/100 ml in the elderly.

An increase in serum cholesterol (hypercholesterolemia) is found in diabetes mellitus, nephrotic syndrome, obstructive jaundice, hypothyroidism, xanthomatosis, atherosclerosis, cirrhosis.

A decrease in serum cholesterol (hypocholesterolaemia) is found in hyperthyroidism, liver disease, pernicious anaemia (except haemorrhagic), malabsorption , malnutrition, oral contraceptive pill.

Procedure:

1. Pipette into labeled test tubes:

Tubes	Blank	Sample	Standard
Sample (supernatant)	-	10 µl	-
Standard (HDL-C)	-	-	10 µl
Reagent	1ml	1ml	1ml

2. Mix and incubate for 5 minutes at $37\,^0$C.
3. Read the absorbance (A) of the sample and standard at 550 nm against the reagent blank within 30 min.

Calculation :

Cholesterol conc. (mg/dL) = (C standard x A sample) / A standard

ESTIMATION OF LOW DENSITY LIPOPROTEIN CHOLESTEROL (LDL–C)

Serum level of LDL-C was calculated according to Friedwald formula, which was based on the assumption that VLDL-C is present in serum at a concentration equal to one fifth (or 0.2) of the Triglyceride concentration. The formula is only valid at triglyceride concentration of less than 400 mg/100ml.

Optimal LDL-C value = 100 –129 mg/dL

Total Cholesterol= HDL–C + LDL-C + VLDL- C

VLDL-C (mg/dL) = 0.2 × Triglyceride (mg/dL)

LDL-C (mg/dL)= Total Cholesterol –[0.2 ×Triglyceride +HDL–C]

Estimation of High Density Lipoprotein Cholesterol (HDL-C)

Principle : Separation method (Precipitant Method) based on the precipitation of apolipoprotein B –containing lipoprotein , Low density lipoprotein (LDL) and very Low density lipoprotein (VLDL) and chylomicron fractions by the addition of phosphotungstic acid in the presence of magnesium chloride ,After centrifugation , the cholesterol concentration in the HDL fraction , which remains in the supernatant , is determined by the usual methods.

Reference normal HDL-C value = 40 – 60 mg /dL

Procedure :

1. Mix 0.2 ml of sample with 0.4 ml of precipitating reagent and incubate for 10 min at 37 °c.
2. Centrifuge for 10 minutes at 4000 r.p.m.,
3. Separate off the clear supernatant within 2 hours.
4. Pipette into labeled tubes:

Tubes	Blank	Sample (supernatant)	Standard
Sample (supernatant)	-	50 µl	-
Standard (HDL-C)	-	-	50 µl
Mono reagent	1ml	1ml	1ml

5. Mix and stand for 10 minutes at room temperature.
6. Read the absorbance (A) of the supernatant and standard at 550 nm against the reagent blank.

Calculation :

HDL-Cholesterol conc. (mg/dL) = C standard x A sample / A standard

15. ESTIMATION OF BLOOD TRIGLYCERIDES

Principle: (Enzymatic Colorimetric Method). The triglycerides are determined after enzymatic hydrolysis with lipases. The indicator is a quinoneimine formed from hydrogen-peroxide, 4-aminophenazone and 4-chlorophenol under the catalytic influence of peroxidase. The red quinine formed is proportional to the amount of TGs present in the sample.

Reference normal TG value < 150 mg /dL.

$$\text{Triglycerides} + H_2O \xrightarrow{\text{Lipase}} \text{Glycerol} + \text{fatty acids.}$$

$$\text{Glycerol} + \text{ATP} \xrightarrow{\text{Glycerol kinase}} \text{Glycerol} - 3 - \text{phosphate} + \text{ADP}$$

$$\text{Glycerol}-3\text{- phosphate} + O_2 \xrightarrow{\text{Glycerol phosphate oxidase}} \text{Dihydroxy acetone phosphate} + H_2O_2$$

$$H_2O_2 + 4\text{-amino-phenazone} + \text{p-chlorophenol} \xrightarrow{\text{Peoxidase}} \text{quinoneimine} + H_2O$$
$$\text{coloured complex}$$

Clinical elucidation:

Elevated levels of triglycerides in plasma have been considered as risk factors related to atherosclerotic diseases. The hyperlipidemias can be inherited or they can be secondary to a variety of disorders of diseases including:

1- Alcohol intake , Obesity

2- Drug intake like (estrogen, B-blocker, thiazide)

3- VonGeirke disease (glucose -6-phosphatase in liver and kidney)

4- Systemic lupus erythromatosus

5- Hormone abnormality (insulin, adrenaline, thyroid, growth hormone)

6- Diabetes mellitus , Chronic renal failure, Biliary obstruction

Procedure :

Pipette into labeled tubes:

Tubes	Blank	Sample	Standard
Sample	-	10 µL	-
Standard	-	-	10 µL
Working reagent	1ml	1ml	1ml

Mix and incubate 5 min at 37 °C .Measure the absorbance of the samples and standard against blank reagent at **550 nm**.

Calculation :

$$\text{Triglycerides Conc. (mg/dL)} = \frac{\text{C standard x A sample}}{\text{A standard}}$$

16. Estimation of Blood Phosphorus (Inorganic)

Phosphorus is present in blood as inorganic phosphate and in combination with several organic compounds including carbohydrates, lipids and nucleotides. Inorganic phosphorus is present mainly in serum.

Method of Fiske and Subbarow

Principle: Serum is deproteinized with iron trichloracetic acid. Protein-free filtrate (supernatant) is mixed with molybdic acid to form molybdic phosphate which is then reduced by 1, 2, 4-aminonaphtholsulphonic acid to produce blue colored molybdenum . The intensity of the color is measured colorimetrically.

Clinical elucidation :

The normal range of serum inorganic phosphorus is 2.5 – 4.5 mg/ dL in adults and 4 – 6 mg/ dL in children.

A-Serum inorganic phosphorus increases in:

1- Hypervitaminosis D,

2- Hypoparathyroidism,

3- Renal Failure

4- During Healing Of Fractures.

B-Serum inorganic phosphorus decreases in:

1-Rickets, osteomalacia, Vitamine D deficiency,

2-Steatorrhoea,

3-Hyperparathyroidism,

4-Disorder of renal tubular reabsorption (fanconi syndrome), renal tubular acidosis.

5-Physiological decrease of phosphorus occur after high carbohydrate diet and after injection of insulin in treatment of diabetes.

17. Estimation of Blood Calcium

Calcium is present in serum in three forms – calcium bound to proteins, calcium bound to inorganic anionic e.g. citrate, phosphate and ionized calcium.

Most of the physiological functions of calcium depend upon the ionized fraction, but for routine work only total calcium in serum is estimated. A convenient and commonly used method is that of Clark and Collip.

Method 1:
Principle: Serum is treated with ammonium oxalate to precipitate calcium as calcium oxalate. The precipitate is washed with ammonia to remove excess oxalate and then treated with sulphuric acid to convert calcium oxalate into oxalic acid. The latter is titrated with standard potassium permanganate.

Method 2:
Principle : Calcium ions form a violet complex with O-Cresophthalein complex-one in an alkaline media . It is a Colorimetric Method using O-Cresophthalein complex, without deproteinization.

Clinical elucidation :

Serum calcium varies from 9 – 11 mg/dL in healthy persons.

1-Serum calcium increases in:
hyperparathyroidism, hypervitaminosis D, multiple myeloma, extensive metastatic involvement of bones, sarcoidosis, idiopathic infantile hypercalcaemia, milk and alkali syndrome, and polycythaemia.

2-Serum calcium decreases in:
hypoparathyroidism, rickets, osteomalacia, steatorrhoea, nephrotic syndrome, renal failure, acute pancreatitis and starvation. In rickets, the serum calcium and phosphorus are decreased .

Procedure :

Pipette into labeled tubes:

Tubes	Blank	Sample	Standard
Sample	-	50 µL	-
Standard	-	-	50 µL
Distilled water	50 µL	-	-
Reagent 1	1ml	1ml	1ml
Reagent 2	1ml	1ml	1ml

Mix and read the absorbance of the samples and standard against blank reagent at 578 nm (550-590 nm) after 5 -50 min.

Calculation :

Calcium conc. = C standard x A sample / A standard

C standard = 10 mg/dL = 2.5 mmol/L

Calcium conc. (mg/dL) = 10 x A sample / A standard

Calcium conc. (mmol/L) = 2.5 x A sample / A standard

Normal serum values = 8.10 - 10.4 mg/dL = 2.02 -2.60 mmol/L

REFERENCES

1. Bishop, M. L., Fody, E. P., and Schoeff, L. (Eds.) (2005). Clinical Chemistry Principles, Procedures, Correlations, 5th ed. Philadelphia: Lippincott Williams and Wilkins.

2. Burtis, C. A., Ashwood, E. R., and Bruns, D. E. (Eds.) (2008). Tietz Fundamentals of Clinical Chemistry, 6th ed. Philadelphia: Saunders.

3. Haven, M. C., Tetrault, G. A., and Schenken, J. R. (Eds.) (1995). Laboratory Instrumentation, 4th ed. New York: Van Nostrand Reinhold.

4. Kaplan, L. A., Pesce, A. J., and Kazmierczak, S. C. (Eds.) (2003). Clinical Chemistry Theory, Analysis, Correlation, 4th ed. St. Louis: Mosby.

5. Westgard, J. O., Quam, E., and Barry, T. (Eds.) (1998). Basic QC Practices. Madison, WI: WesTgard® Quality Corporation.

Printed by Books on Demand GmbH, Norderstedt / Germany